POWER UP WITH ENERGY!

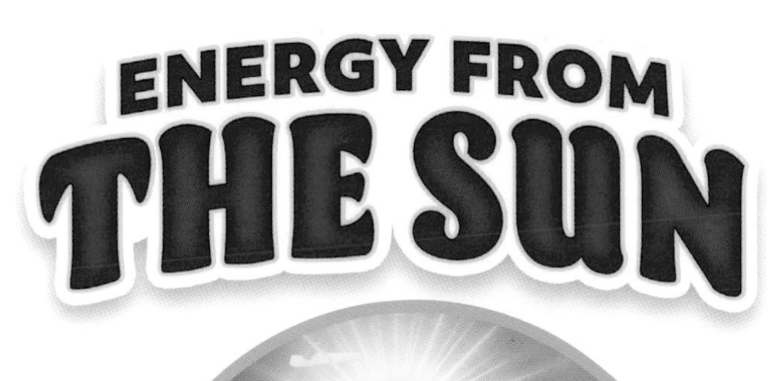

Energy from the Sun

by Karen Latchana Kenney

Consultant: Beth Gambro
Reading Specialist, Yorkville, Illinois

Minneapolis, Minnesota

Teaching Tips

Before Reading

- Look at the cover of the book. Discuss the picture and the title.
- Ask readers to brainstorm a list of what they already know about the sun. What can they expect to see in this book?
- Go on a picture walk, looking through the pictures to discuss vocabulary and make predictions about the text.

During Reading

- Read for purpose. Encourage readers to think about the sun and energy and the roles they play in our daily lives as they are reading.
- Ask readers to look for the details of the book. What are they learning about the sun?
- If readers encounter an unknown word, ask them to look at the sounds in the word. Then, ask them to look at the rest of the page. Are there any clues to help them understand?

After Reading

- Encourage readers to pick a buddy and reread the book together.
- Ask readers to name one reason to use energy from the sun and one reason to not use it. Go back and find the pages that tell about these things.
- Ask readers to write or draw something they learned about energy from the sun.

Credits:
Cover and title page, © Smitt/iStock, © stocktech78/Shutterstock; 3, © tma1/iStock; 6–7, © bruev/iStock; 8–9, © hadzi3/iStock; 10–11, © CL Shebley/Shutterstock; 12–13, © VioNettaStock/iSTock, © timsa/iStock; 14, © mrfotos/Shutterstock; 15, © Diy13/iStock; 16, © Smileus/iStock; 17, © varuna/Shutterstock; 18–19, © ofc pictures/iStock; 20–21, © gorodenkoff/iStock; 22, © MoreISO/iStock, © Madmaxer/iStock, © Mi Pan/Shutterstock; 23BL, © Ryhor Bruyeu/iStock; 23BR, © Smitt/iStock; 23TL, © Alberto Masnovo/iStock; 23TR, © Dmitry Galaganov/Shutterstock

Library of Congress Cataloging-in-Publication Data

Names: Kenney, Karen Latchana, author.
Title: Energy from the sun / by Karen Latchana Kenney, Consultant Beth
Gambro, Reading Specialist, Yorkville, Illinois.
Description: Minneapolis, Minnesota : Bearport Publishing Company, [2022] |
Series: Power up with energy! | Includes bibliographical references and
index.
Identifiers: LCCN 2021001068 (print) | LCCN 2021001069 (ebook) | ISBN
9781647478681 (library binding) | ISBN 9781647478759 (paperback) | ISBN
9781647478827 (ebook)
Subjects: LCSH: Solar energy--Juvenile literature.
Classification: LCC TJ810.3 .K458 2022 (print) | LCC TJ810.3 (ebook) |
DDC 621.47--dc23
LC record available at https://lccn.loc.gov/2021001068
LC ebook record available at https://lccn.loc.gov/2021001069

For more information, write to Bearport Publishing, 5357 Penn Avenue South, Minneapolis, MN 55419.
Printed in the United States of America

Contents

Light up the Night

It is getting too dark to play.

But then the yard lights turn on.

What makes them work?

It may be the sun!

Our yard lights need **energy** to work.

It is energy that gives them power.

We can get energy from the sun.

7

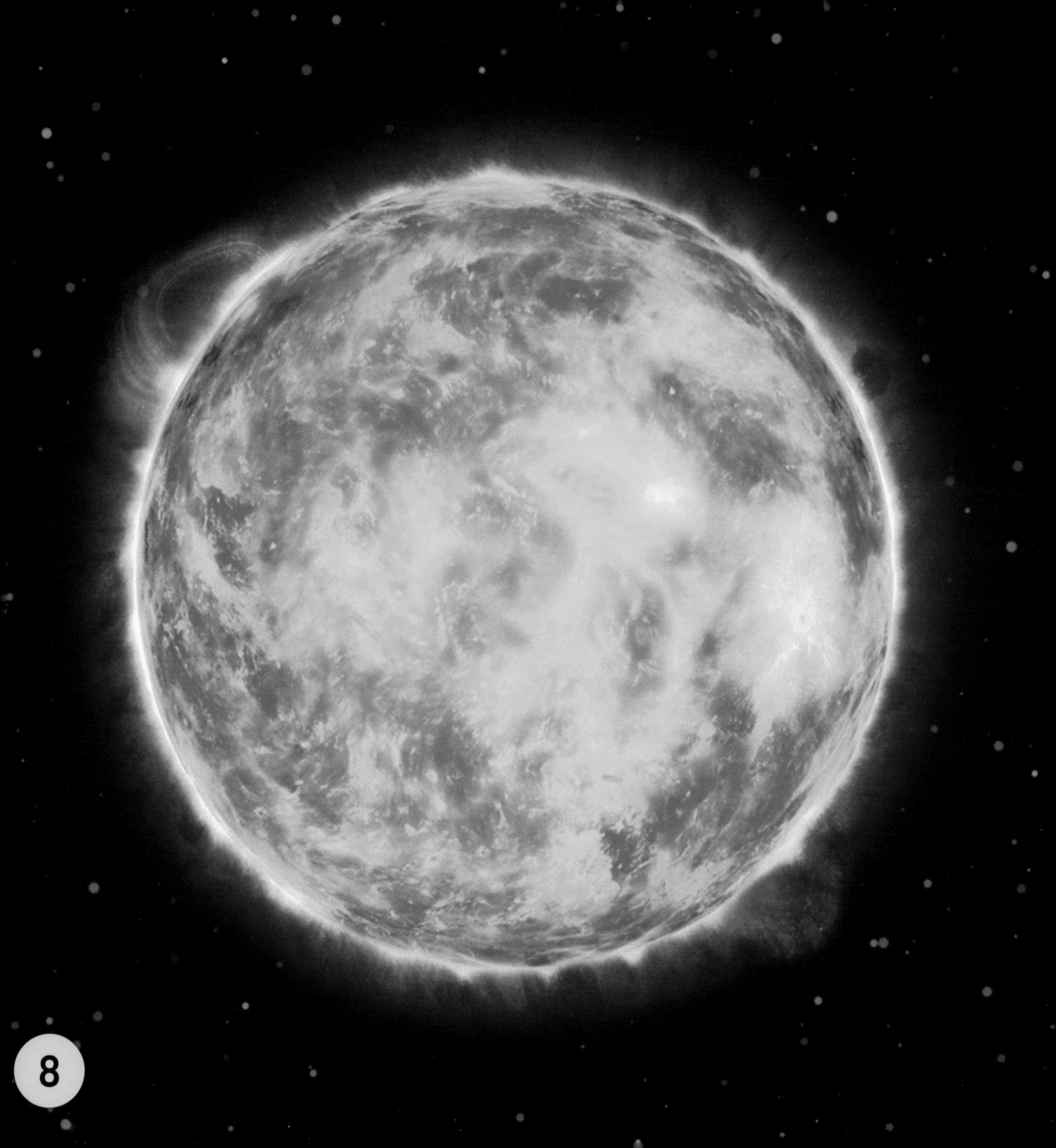
8

We see the sun's energy as light.

We can feel it as heat.

The energy moves through space to Earth.

During the day, the sun's energy is all around.

At night, we can use it, too.

But first we need to do something to it.

A panel

Light from the sun lands on flat **panels**.

Small **cells** on the panels take in light.

Then, they turn light into power.

A cell

Later, we can
use the power.

Some traffic lights
use the sun's power.

Homes and buildings use
it, too.

It can even **charge** phones.

The sun is out every day.

Each day we can use it for energy.

It makes power that does not hurt Earth.

Some things about getting power from the sun are hard.

At night, we cannot make power from the sun's energy.

And panels can cost a lot.

Even so, new panels are put up all the time.

Energy from the sun can be good for us all.

Soon, more people may be able to use it.

Energy from the Sun

Follow along as light from the sun becomes power.

1. Light comes from the sun.

2. The light hits a panel.

3. Cells on the panel change light into power.

4. The power can make machines work.

Glossary

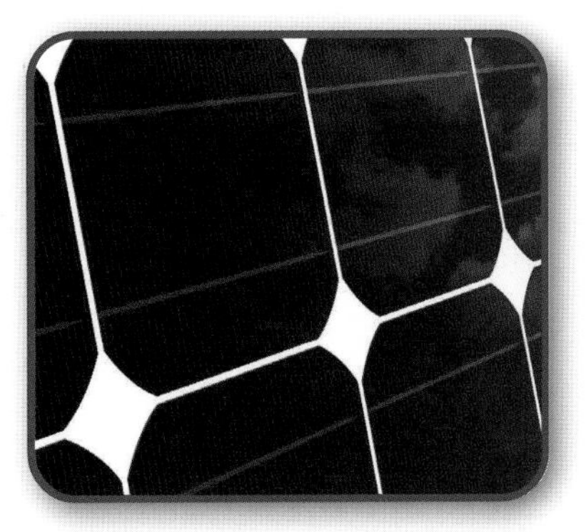

cells basic, very tiny parts of things

charge to give power to something

energy power that makes things work

panels large, flat pieces

Index

Read More

Felix, Rebecca. *Solar Energy (Earth's Energy Resources).* Minneapolis: Abdo, 2019.

Schuh, Mari. *Where Does Light Come From? (Let's Look at Light).* North Mankato, MN: Capstone, 2020.

Learn More Online

1. Go to **www.factsurfer.com**
2. Enter "**Sun Energy**" into the search box.
3. Click on the cover of this book to see a list of websites.

About the Author

Karen Latchana Kenney likes biking and reading. She tries to find ways to use less energy every day.